organic basics

soil

Charlie Ryrie

NEWTON LE WILLOWS
LIBRARY
TEL 01744 677885 , 6 , 7

Gaia Books Ltd

CONTENTS

4 Preface

6 **Why** worry about soil?
8 Why is my soil so important?

10 **What** is my soil?
12 What's in my soil?
14 What do the layers mean?
16 Identifying my soil
18 Soil structure and fertility
20 What are the major nutrients?
22 Why does pH matter?
24 Plants for particular soils
26 Soil basics – checklists

28 **How** do I improve my soil?
30 How and when should I dig?
32 Digging methods
34 Tools for the job
36 Cultivating without digging
38 The magic of compost
40 Using manure
42 How and when to mulch
44 More soil improvers
46 Liquid fertilisers
48 Green manures
50 Drainage difficulties
52 Rotation
54 Beginning in beds
56 Raised beds
58 Soil in containers

60 Resources
61 Index

PREFACE

Soil Association

The membership charity
campaigning for an organic Britain

It is hard to imagine but 1 gram of healthy soil can contain more than one billion organisms of over 10,000 different species. This vastly complex and diverse ecosystem has suffered over the last fifty years due to the intensification of agricultural practice. The soil has in many situations come to be regarded simply as an anchor for crop roots rather than the driving force behind an ecologically balanced food production system. Lady Eve Balfour, the founder of the Soil Association, proclaimed the importance of the soil at a time when agrochemical sales were beginning to escalate globally.

'The health of man, beast, plant and soil is one indivisible whole; the health of the soil depends on maintaining its biological balance and, starting with a truly fertile soil, the crops grown on it, the livestock fed on those crops and the humans fed on both have a standard of health and power of resisting disease and infection greatly in advance of anything ordinarily found in this country.'
Lady Eve Balfour, Living Earth, 1943.

This link between soil health, crop health and human health remains at the heart of organic farming today. The first step for gardeners must be to learn and understand how their soils behave and what practices they can use to protect and enhance soil life and soil structure. This book covers these practices in a way that can be practically applied in a broad range of situations. I can guarantee that an understanding of the workings of the world beneath your feet will revolutionise your approach to growing vegetables.

Rob Hayward
Horticultural Development Officer
The Soil Association

Civilizations which have neglected their soil have perished. Soil is essential to life, and has a complex ecology of its own which is as yet poorly understood. Although each teaspoon of soil contains a myriad of soil-dwelling organisms, modern agriculture, and to some extent gardening, tend to treat it as a sterile growing medium which can be given a quick fix with fast-food, chemical nutrients.

Organic gardeners feed their soil with wholefood – slow-release garden compost and well-rotted manures. This wholefood diet ensures that all the soil creatures, from the microscopic bacterium to the six-inch-long earthworm, can play their full role in producing healthy crops.

Feeding is only part of the story. Gardeners need to get to know their soil so that they can work with it, rather than against it. This doesn't just mean doing a pH test – it means understanding the structure and characteristics of the soil type. All soils have their positive and negative points, but they can all be improved with organic matter.

Our knowledge of the soil is in its infancy, but we can use the knowledge we have to produce better crops without the use of chemical fertilisers.

HDRA, the organic organisation, is one of the oldest environmental organisations. It researches, promotes and advises on organic cultivation for home gardeners everywhere. It works to continually improve composting, pest and disease control and other techniques, to enable organic gardeners to produce better crops of sturdy, healthy vegetables, fruits and ornamentals.

Judy Steele
Editor, Organic Way
HDRA, the organic organisation

the organic
organisation

An international membership
organisation, researching and
promoting organic horticulture
and agriculture

WHY

WORRY ABOUT SOIL?

Start with the soil.
You can't grow healthy
plants in undernourished
soil. But once you get your
soil right you'll have no
problems growing what you
want, where you want it.

WHY IS MY SOIL SO IMPORTANT?

There is no mystery to gardening. There are no such people as green-fingered gardeners. But like everything else there are certain basic rules that are simple once you understand how they work, puzzling if you don't.

When you take on a garden it is a natural reaction to want to get planting as soon as possible, to make your mark. But the best gardens are not created in an instant, they develop more slowly. Organic gardeners work with nature, so you should learn to think about what your garden wants from you as well as what you want from it.

If you buy a plant from a nursery, dig a hole in unprepared soil and put it straight in, it may grow. It probably will survive for one season anyway. But then it will run into problems. You could simply blame the plant, or the supplier, and try again, perhaps feeding the plant at some stage. You will probably still have limited success. If you put another exactly similar plant into a bed of well-nourished soil it will flourish. This isn't magic, just sound common sense. Soil is the basic building block in our gardens. If you get that right, everything else will follow.

Plants need many of the same things as humans: they need air, water and food. The soil needs to supply all these needs, providing nutrition in forms that plants can use. Plants feed through their roots which need to be able to delve and spread through the soil. If it is fertile – in good condition and full of nutrition – you'll have contented plants; if not, they'll have a hard time. People can survive for a while on junk food but they won't be healthy. It's the same with plants: if you give them rubbish soil they are stressed and can't reach their potential.

What's going on?
There is much more to soil than immediately meets the eye. It is a heaving world of different life forms and minerals, a mixture of air, water, weathered rocks and organic matter, broken down by the actions of billions of living

creatures and tiny organisms. Healthy soil is a sort of natural chemical factory teeming with life, where millions of micro-organisms convert minerals and chemicals into substances plants can use to grow. You don't need to know the details of how all these processes work (unless you want to) but you do need to recognise that they are all important and interrelated, and to provide the right environment for them to take place. It's not difficult to keep your soil healthy, and it is entirely rewarding.

Take a look at a productive plot. On the surface you see lush foliage, abundant fruit, probably a few weeds. Now think what is going on beneath the surface to support all that exuberant growth. Think of all the roots tunnelling down to get at the nutrients they need. Imagine all the tiny creatures working away to transform decaying matter and newly released minerals from crumbling rocks into useful food. There's a lot happening. Once you can visualise what's going on, it's obvious how difficult it must be for all these processes to go on in a very dry, thin, hard soil, or if it is solid, wet and airless; how much easier everything must be if the soil is deep, moist and crumbly.

Keeping it going

If your soil is going to support plant life year after year you must look after it. Otherwise it will eventually support little or nothing at all. This is what happens to land fed year on year with artificial fertilisers – all its own natural goodness is used up and fertility comes purely through artificial additives. The soil becomes useful only to keep plants from toppling over and to provide water.

But if you look after your soil organically, it will reward you with flourishing plants that attract beautiful wildlife and make your garden a glorious place to be. As you are working with nature there will always be plenty of suprises. But most of your gardening adventures will be successful ones if you get to know your soil, and work with it.

WHAT

IS MY SOIL?

Soil is alive, a mineral, animal and vegetable base teeming with organisms that work together to keep everything in the optimum condition needed to support healthy plants. Once you understand what you're dealing with, keeping it in good condition becomes common sense.

WHAT'S IN MY SOIL?

All soil – except for peat, which is a special case – is made up of three layers of rock in different stages of decomposition, although these layers are sometimes rather indistinct.

Topsoil is dark brown and crumbly, full of organic matter and teeming with life. Most plant activity takes place here.

Below topsoil is subsoil. This is made up mainly of rock particles and is usually paler than topsoil because it contains little organic matter. Here there is less plant activity, although deep rooted plants mine the subsoil for nutrients. Subsoil contains few living organisms, but much of the water that plants need is stored here, along with important minerals waiting to be made available to growing plants.

Beneath subsoil is bedrock, the basis of all soil.

It has been calculated that a single earthworm can shift 30 tonnes of earth in its lifetime!

You may never have thought much about exactly what is in your soil, but once you know a little about how it is made up and what goes on, you should be able to make sure that it always contains a good balance of plant food, in conditions where plants can use it. If you get that right, your garden will always flourish.

Our soil has been formed over thousands of years from the gradual breakdown of rocks in the bedrock and sub-soil, through their fragmentation and erosion by water, ice, heat and cold, and the chemical changes they have been through. But it is far from an inert mass of bits of weathered rock. It also contains decaying matter from plants and animal wastes that fall onto the soil. It con-tains air and water, and a seething mass of living insects, grubs and micro-organisms that convert the minerals, organic matter, air and water into nutrition for plants, and ultimately for animals and humans.

Plant food

The bulk of soil is tiny crumbs of minerals. How they are distributed depends on climate and cultivation as well as the underlying rock formation. Plants absorb those they need through their roots in solution from the water in the soil. Mixed with the mineral particles is other plant-food-to-be in the form of remains of dead plants and animal wastes, or organic matter.

Any organic matter that falls on the surface of soil will eventually be incorporated into it as it gradually weathers and as living creatures and tiny organisms break it down and combine it into the soil. Worms and insects drag rot-ting material down, digest and excrete it and begin the process of transforming it into humus. Soil bacteria, fungi and algae then release the nutrients from humus in a form plants can use. So all organic matter is transformed in the soil into more useable minerals, plus the proteins, carbohydrates and sugars that plants and soil organisms need. Then plants grow and feed insects and animals, falling plant matter and manure is deposited onto the soil again and the cycle of life continues.

Air and water

Soil can only do its job of supporting plant and animal life if it contains sufficient oxygen. Plants need oxygen just as much as humans. They take it in through their roots and give out carbon dioxide. The living organisms in the soil also need oxygen to speed up the decaying processes that turn organic matter into nutrition. Oxygen gets into the soil from the air via the passages made by burrowing creatures, through the spaces between crumbs of soil and from the air brought in with water draining down through the soil.

The nearer the air to the surface of the soil, the greater its oxygen content; at lower levels soil air contains more carbon dioxide. If soil is not sufficiently aerated it contains an unhealthy balance of carbon dioxide to oxygen which can poison plant roots and living organisms.

The water content of soil is equally crucial. Plants get food in solution through their roots. If there isn't enough soil water they can't absorb necessary nutrients and starve. Thirsty plants can't make efficient use of sunlight, and photosynthesis can't proceed. But if they get too much they they can't absorb the air they need, so they drown and rot.

Helping nature

In nature plants adapt to their particular conditions and different plants thrive in different types of soils, but if you understand how soil works you can help nature to overcome most restrictions. Even if your soil starts off thin, stony and dusty, or heavy, wet and muddy, with time and good practice you will be able to build its fertility and grow the widest range of plants possible in your climate.

Peat soils are in a class of their own. They are formed not from rock but from vegetable matter which has been compressed under water and has not rotted away. Naturally drained peatlands are the most fertile soils in the world.

Not all the living creatures in soil are helpful ones. Most soils also contain some pests and disease organisms, but once you get to know how your soil works you should be able to manage it so you have little trouble from pests or diseases.

WHAT DO THE LAYERS MEAN?

Hardpan is very common in gardens
on new housing developments.
Whereas you should generally leave
subsoil well alone, and concentrate
on cultivating the topsoil above, you
will need to break up any hardpan
as it forms an impervious layer
preventing drainage or nutrients
getting to the plants.

Lime facts
- Ground limestone is easy to apply and widely available
- Slaked lime is limestone which has been burnt then moistened with water
- Sea sand is rich in calcium from the crushed shells of molluscs

The good thing about soil is that, whatever you start with, you will probably be able to turn it into a fertile growing medium. The more you know about what you are dealing with, the more likely that is.

Dig a hole at least half a metre deep and look at the depth and colour of the different layers. In general the deeper the topsoil, the better. It's best if your topsoil is well over a spade's depth (25cm) or you will be constantly bringing up subsoil when you dig. If your topsoil layer is shallow you need to concentrate on building it up with copious additions of organic matter, otherwise you will fight a constant battle to maintain the moisture and nutrients your plants need.

In some areas you may get down to bedrock at well under a metre, in which case you need to think about bringing in more topsoil as it is hard to change the overall depth of soil within one gardener's lifetime.

In some situations it is hard to build up the topsoil – for example, on a steep slope where water just runs off. You may need to build raised beds (see page 56) or restrict your plant selection – if you aim for deep-rooted or slow-growing vegetables and a lush herbaceous border, then gardening will never be a pleasure but a chore.

Subsoil colour

Looking at the subsoil gives you a good idea of your soil's good points and problem areas. Blue-grey subsoil indicates lack of oxygen and therefore poor drainage. On the other hand reddish or yellow subsoil indicates high levels of oxidation from the action of air on the weathering rocks. If air is getting through the soil it is also likely to have good drainage. However, if there appears to be a band of red or brown subsoil up to half a metre below the surface this is probably hardpan, a layer of compacted soil. This is usually caused by the action of heavy machinery on the soil surface, or sometimes by bad drainage.

IDENTIFYING MY SOIL

Clay particles are much too small to see at around .0002mm diameter. Sand particles range from 0.2-2mm diameter. Anything larger is gravel.

Even an area of predominantly sandy soil may have clay or loam pockets, or vice versa, depending on the underlying geology. Soil in different areas of your garden may also behave differently depending on the levels of cultivation.

Green-fingered gardeners may not exist, but some soils certainly present more of a challenge. The first step when you take on a garden is to identify the type of soil you'll be working with.

Soil texture

When we talk about sand, clay or loam soils we are refer-ring to the size of the mineral particles, or the soil's texture. Sandy soils have large particles, which means they don't bind together closely and there are significant spaces between the particles. They are sometimes referred to as light soils and they are easy to work, but hard to keep in good condition as they tend to drain fast.

Clay soils are made up of tiny mineral particles which stick together like glue and provide a large surface area. Clay soils tend to be solid and heavy to work, and difficult for air and water to penetrate, so they can get stale or waterlogged, but they hold nutrients well. Silt soils have slightly larger particles than clay and are easier to work, but get sticky and cold when wet and dry into dust.

Loam soils are made of a mixture of small and large particles. They are usually divided into sandy loams and clay loams, depending on the proportion of clay particles they contain. Peat soils also contain a mixture of small and large particles.

Identity check

Some gardeners say they can tell everything they need to know about their soil by smelling it or tasting it, but the easiest check is just to pick up a small handful and try to roll it into a ball in the palm of your hand.

If it feels gritty and refuses to stick together it is sand. If it is slightly gritty but forms a soft dark ball it is loam, and a soft springy dark ball is peat. If it makes a sticky ball, it is clay or silt but if you can make the surface shine when you rub your thumb over it, you're holding clay.

SOIL STRUCTURE AND FERTILITY

Examining soil structure

As the structure of your soil is so crucial it is worth examining it quite closely. The best way is to carry out some simple tests. Dig a hole about 25cm deep, and see whether the soil is a fairly even texture or if there are solid patches. A well-structured soil should be very crumbly at and near the surface and some crumbs should be visible even at a depth of 25cm. Look at how the soil removed from the hole hangs together. It should be dark with plenty of 1cm crumbs and some larger clods that break apart quite easily in the hand. If it is pale or streaky in colour, rather sticky and dense, or gritty and dusty, its structure will need attention.

Try dividing a handful of your soil between two glasses, then gently pour water over one of the samples. If the structure of your soil is good, the crumbs will hold together even when they're wet so the wet and dry samples should look the same. Or mimic the action of heavy rain and pour a jug of water onto bare soil in your garden. This may flatten all the soil crumbs into a fine even layer, an effect known as puddling, in which case the structure needs improving.

It is a myth that one type of soil is always more fertile than another, although some types are easier to work than others. A fertile soil provides all the nutrients plants need by maintaining the right environment for all the living organisms and micro-organisms to convert minerals and organic matter into plant nutrition.

This fertility depends principally on the structure of the soil, or the size of the crumbs which join the lumps of mineral and organic particles together. The size of the spaces around the crumbs is just as important, as the air and water that plants and soil life need get into the soil through these spaces. In an ideal soil the soil crumbs and spaces will vary in size.

The structure of the soil determines the way it converts and stores food for plants. If it's too dense, air and water won't be able to enter it evenly, organisms won't be able to do their job, and roots won't be able to get at the food they need. If the structure is too loose water will drain through the topsoil, leaching nutrients with it.

As a guideline, soil needs to be crumbly enough for worms to be able to move through it freely, digging, dragging and burying organic matter, and it needs to be spongy enough to contain enough moisture and air for plants to get at all the food they need. Whatever you start with, you can end up with soil where plants can flourish.

Organic matter

A poorly structured soil can't be improved overnight, but if you follow one of organic gardening's mantras and 'add organic matter' you'll be on the way.

Organic matter is usually taken to mean manure or compost, but it is nothing more mysterious than material that was once alive, from fallen leaves to cotton socks, dead insects to bales of straw. Most organic matter in the garden is plant debris and animal manures plus remains of small creatures and tiny organisms. It is incredibly important not only because it provides a balanced diet for your soil, and therefore your plants, but because it improves the structure so that plants can benefit from the nutrition. It helps soil particles to stick together in crumbs, enabling large sandy particles to hang together a bit more, and helping to break up the hold tiny clay particles have on one another and encouraging them to join with larger particles.

Strategies for improvement

Loams are the easiest soils to keep in fertile condition as their crumbs are naturally different sizes so water and nutrients are easily absorbed and loam drains well.

If you have a heavy clay-based soil you will need to open it up so that air and water can penetrate, and to improve its drainage. If your soil is light and sandy you need to make it stick together more so that water does not run away.

It takes time and can be hard work to change the structure of a compacted clay soil. You need to dig in grit to improve aeration, and add well rotted organic matter to improve the soil life. It is impossible to give precise quantities of grit as everyone's soil is different, but it is hard to add too much if you're adding plenty of organic matter as well. Make sure that the compost or muck you add to clay soil really is well rotted otherwise you can compound problems. An unimproved clay soil will have poor soil life because of problems with air and water, so it may not contain enough organisms to deal with tough semi-rotted organic material.

Sandy soil needs to be made less gritty and more sponge-like, by adding copious amounts of organic matter at regular intervals. Because sandy soils contain a lot of air anything you put into them will oxidise and rot quickly, so you can add muck, compost or other matter in earlier stages of decay.

Plants and soil life suffer in soil that is too wet or too dry.

Never work soil when it is wet, as digging or treading makes the structure even worse.

Try not to leave soil uncovered as this can harm the structure, causing wind and rain erosion in light soils and puddling and waterlogging in heavy soils. Ideally, mulch well (see page 42) or sow green manure (see page 48).

WHAT ARE THE MAJOR NUTRIENTS?

Nitrogen – N
Vital part of all protein, essential for leaf and stem growth. Demand for nitrogen is strongest when new plant tissue is developing.

Phosphorus – P
Vital for cell division, plant maturation, and root development.

Potassium – K
Vital for flowering and fruiting, to strengthen plant tissue and help water absorption.

• Nitrogen deficiences usually show as sickly yellow plants (see p.27). Solve by regular addition of compost and muck.

• Clay soils are usually potassium-rich but if you suspect a deficiency (see p.27) you may need a potash-rich additive such as comfrey.

• If your soil lacks phosphorus (see p.27), you can dress it with rock dust, or rock phosphate. This is inert, can be added at any time of year, won't harm the plants and will continue to benefit the soil for many years.

Plants get the carbon they need from carbon dioxide in the air, but all the other nutrients they need are supplied from the soil, which needs to contain a balanced spread. The most important plant nutrients are known as the major elements, or macronutrients, because plants need them in relatively large quantities. These are nitrogen (N), phosphorus (P) and potassium (K). You may hear gardeners talking about the NPK balance of their soil, and if you buy an organic fertiliser it will indicate the percentages of NPK it contains. Calcium, magnesium and sulphur are also major players.

Then plants need trace elements in smaller quantities, including iron, magnesium, boron, zinc, copper and molybdenum.

These elements occur naturally from the weathering of mineral particles in the soil, and they are slowly released as organic matter decays. Nitrogen also comes from the air in the soil, converted into a form plants can use by bacteria, including nitrogen-fixing bacteria on the root nodules of peas and beans. As adequate nitrogen is vital for strong and leafy growth, nitrogen has the most obvious effect on your plants. The level of calcium, or pH level (see page 22), is also vital as this can affect the ability of plants to access all the other elements in your soil.

Keeping the balance
If your soil has a good structure and you add farmyard muck, leafmould and compost in a three- or four-year cycle, it should maintain an adequate balance of all the elements. Manure is nitrogen-rich and garden compost is an ideal balanced source of nutrients, providing major and minor elements in a form plants can use. If plants still do not thrive, the soil could be lacking some essential mineral. You should get your soil analysed if you have persistent problems, but common deficiencies are easily recognised (see page 27).

WHY DOES PH MATTER?

The balance of minerals in most soils can be improved through regular cultivation. Test your soil every two or three years if you have any doubts about its fertility.

Never add slaked lime or hydrated lime as these are very soluble and quick acting so you shouldn't use them in an organic garden. The best lime comes from Dolomitic limestone as this contains significant magnesium as well as calcium.

Do not lime an area of the garden where you intend to grow tomatoes or potatoes the following season, as they are sensitive to excessive lime. Brassicas, on the other hand, love recently limed soil.

In very alkaline soils plants may be unable to get at the micro-nutrients they need. Iron, copper, and manganese, for example, become locked up in insoluble compounds in very alkaline soil.

If vegetable growth in your garden is poor, or soil is moss-covered with a rather stagnant look, or you have lots of weed growth, particularly sorrel and docks, you should purchase a kit to check the pH of your soil. This indicates whether your soil contains too little or too much calcium, or how acid or alkaline it is. pH is measured on a numerical scale. The neutral point on the scale is 7; soils below that become increasingly acid (too little calcium) and above 7 are increasingly alkaline (too much) .

The most fertile soil for growing vegetables is very slightly acid, around pH 6.5, but from just below 6 to just above 7 is fine. Most trees and shrubs and herbaceous plants will also grow happily in soil with this pH but some fruits prefer slightly more acid conditions.

If the calcium content is off balance, this can affect the availability of other elements in the soil – in very acid or very alkaline soils plant roots have difficulty in extracting whatever minerals may be available to them. It can also harm or deter the soil organisms that keep your soil healthy, so it is important to get it right.

Testing the soil

It's a good idea to take samples from different parts of your garden and see what the difference is. If the pH in a cultivated area is fine, but it is too acid or alkaline elsewhere, regular cultivation and addition of organic material may be all your soil needs. In some gardens there may be pockets within quite a small area which are considerably more acid or alkaline, depending on the underlying geology as well as how much the soil has been worked, so regular testing is an excellent way of getting to know the soil profile of your garden.

The usual way to correct an acid soil is to add ground lime to the soil. Lime also improves the structure of clay soils by helping the particles to stick together in crumbs, and it inhibits bacteria that remove nitrogen from the soil. But it's not the only solution. One problem with lime is that earthworms don't appreciate it, so it may take a while for soil life to build up again after liming.

Acid soils

Soils in cooler damp climates such as the UK tend to become increasingly acid, even if well cultivated, as rain is constantly washing calcium out of the soil. If your soil is very acid, aim to bring the pH up slowly, over three or more seasons. Overliming causes its own problems, inhibiting plant growth by suppressing the actions of some important trace elements in the soil. On sandy soil, dress the surface of the soil with around 200g lime per sq. metre; loamy soil will take more, and heavy clay more still.

Wait several months before applying muck to an area you have limed as the reaction between lime and manure will cause the escape of ammonia as a gas, wasting nitrogen that could otherwise be used by the soil. Apply ground limestone to the surface of the soil in autumn, before cultivating in spring.

Alkaline soils

It is rare for cultivated soils to be over-alkaline, as regular applications of muck and compost will usually supply all the elements soil needs, apart from lime. Most vegetables will in any case grow in slightly alkaline soil.

If you can't lower your soil's alkaline level with regular cultivation, you can apply gypsum or powdered sulphur at the rate of 130gm per sq. metre. You will be more guaranteed to succeed in creating fertile growing conditions if you build raised beds (see page 56) or dig some trenches and fill them with imported acid soil to get plants going.

pH scale

· pH 4-5
Acid. Found in cold wet areas.
Camellias and rhododendrons thrive,
also blueberries and cranberries.
Few earthworms, little soil life.

· pH 5-6
Fairly acid. Typical of unimproved
soil in very wet areas.
Potatoes, tomatoes, and most fruits
thrive.
Soil life survives.

· pH 6-7
Neutral.
Most plants and garden crops thrive.

· pH 7-7.5
Alkaline. Typical of hot dry areas.
Most garden plants survive.

· pH above 8
Very alkaline, typical of semi-desert
areas.
Soil life and plants struggle.

PLANTS FOR PARTICULAR SOILS

Plants for acid soils
bilberry
cranberry
rhubarb
strawberry
raspberry
tomato
celery
potato
parsnip
red maple
rhododendron
camellia
azaleas
heathers
magnolia
spruce

Plants for alkaline soils
brassica
asparagus
spinach
currant
apple
peach
plum
orchid
fritillary
tulip
holly
beech
cedar
cypress
wisteria
lilac

Plants for sandy soils

onion
carrot
parsnip
beetroot
salad greens
rocket
chard
runner bean
tomato
nasturtium
soft fruit *except strawberries and raspberries*
apple
nut trees
sweet chestnut
holm oak
catmint

Plants for clay soils

pea
broad bean
potato
parsley
squash
strawberry
raspberry
pear
plum
rose
willow
oak
black walnut
poplar
dogwood
elder

SOIL BASICS — CHECKLISTS

Soil properties
- **aeration**
  Sand: good
  Loam: medium
  Clay: poor
  Silt: medium
  Peat: good

- **holding nutrients**
  Sand: poor
  Loam: good
  Clay: good
  Silt: good
  Peat: good

- **waterholding**
  Sand: poor
  Loam: medium
  Clay: good
  Silt: good
  Peat: good

- **ease of working**
  Sand: good
  Loam: good
  Clay: poor
  Silt: good
  Peat: good

Sand
Structure: very open, unwilling to form clods. Needs continuous supplies of organic matter. Never leave sandy soil bare but protect with mulch or green manure when not in cultivation.

Drainage: swift to drain, but nutrients leach out easily and the soil dries out easily and suffers from drought. Organic matter improves drainage by making the soil more spongy.

Temperature: warms up quickly so ideal for early crops.

Cultivation: sandy soil can be worked at virtually any time as long as organic matter is always incorporated.

pH: often chalky and alkaline but check every two years as lime is easily washed out of light sandy soils.

Clay
Structure: tiny particles stick together closely and it needs opening up. Requires grit and well-rotted organic matter.

Drainage: tends to waterlog causing lack of aeration and stressed plants and soil life. However, holds nutrients well. If it dries out it becomes cement-like and impenetrable, so mulch well in summer.

Temperature: cold, not suitable for early crops; because of the high moisture content it is quick to freeze.

Cultivation: only ever cultivate clay soil in dry weather. Add very well rotted organic matter as lack of air means slow decomposition. Turn soil on seed beds in autumn and leave frost to break up the structure.

pH: clay soil tends to be slightly acid; liming encourages clay particles to clump together in larger groups.

Loam

Structure: good structure which is easy to maintain with regular addition of organic matter.
Drainage: drains well but retains enough moisture to hold nutrients. Not drought susceptible.
Temperature: warms up quite early and slow to freeze so provides the longest growing seasons.
Cultivation: do not work in wet, otherwise tolerant.
pH: usually around neutral but check every five years.

Peat

Structure: good.
Drainage: the most fertile peat soils drain well but can dry out into a hard crust; others hold too much water.
Temperature: quick to warm; slow to freeze.
Cultivation: any time.
pH: usually acid, sometimes very acid.

Silt

Structure: good.
Drainage: silt is a combination of sandy and clay deposits and has some of the characteristics of each. It can pack down and become waterlogged, but it also dries out to a free-draining dust.
Temperature: slow to warm, or to freeze.
Cultivation: cultivate when soil is moist, neither wet or dry
pH: typically neutral to acid.

Soil deficiencies

· nitrogen
Symptoms: pale yellowing foliage, small leaves, stunted growth.
Cure: Add muck and compost, improve drainage, rotate crops.

· phosphorus
Symptoms: small pale leaves, slow growth; leaves have blue-green tinge and fall early.
Cure: Add compost, improve drainage.

· potassium
Symptoms: pale leaves and stunted. growth, brown tipped leaves with yellow between veins.
Cure: Comfrey, rock dust.

· magnesium
Symptoms: green veined mature leaves with yellow between veins.
Cure: Epsom salts, compost.

· iron
Symptoms: young leaves yellow between veins.
Cure: reduce pH with liberal compost, or sulphur or gypsum.

· manganese
Symptoms: mature leaves fade.
Cure: raise pH – add lime or chalk.

HOW

DO I IMPROVE MY SOIL?

Learn the best ways to
cultivate your soil, what
organic fertilisers to add,
and how and when to add
them. It's not all hard work.
Organic methods mean that
you are working with nature,
not against it, and you can
even choose to build up a
healthy soil without
touching a spade.

HOW AND WHEN SHOULD I DIG?

You can get an idea of your soil's fertility by checking out the worm population. Turn over a section of soil about 45cm in length and a spade's depth. You should bring up a selection of worms of different sizes. A really fertile soil that has been regularly fed with muck and compost will probably provide about 30 worms in that small area, about half would be a good start. You won't find many worms in a heavy compacted clay soil as it's too hard for them to move through and they can drown in waterlogged soil. You probably won't find many in light sandy soil either, as worms are very susceptible to drought.

Whatever plans you have for your garden, they will all succeed if you get your soil in good condition. But this need not be so much work as it sounds.

Most people assume that digging is an integral part of gardening. But in a properly managed organic garden you should only need to dig your ground once, right at the beginning. Then you need never lift a spade again if you don't wish to. If you are patient, you don't even need to dig at the beginning but can cover your ground with a long-term mulch (see page 36) to clear the ground while you get on with other tasks.

However, most gardeners are itching to get at the soil when they take over a garden, to start the work of improving it and creating a fertile patch as soon as possible. You will want to check the condition of the soil and see where improvements need to be made, and to clear unwanted growth from areas you want to plant. Take as much care as possible to clear the ground thoroughly at this stage, as you do not want to be forced to dig in following years, but you do want to give future plants the best chances of getting at all the available nutrients they need rather than competing with others.

Digging advantages

The prime purpose of digging is to incorporate organic matter along with the air the soil needs for plant roots and soil organisms to breathe. Digging also loosens soils, breaking up heavy clods to allow roots easier penetration, and providing channels for rain water to soak in to the ground. This is very important in a garden that has been neglected, and in gardens on new building developments where the soil is often very poor quality topsoil imported and dumped over hardpan. In such conditions you can add organic matter or other soil improvers (see page 44) when you dig.

Digging disadvantages

While soil benefits from a good shake up in the right conditions, it doesn't appreciate being trodden on as this compacts it. So if you dig too often, or at the wrong time, you will harm soil structure and soil organisms. Clay needs to be opened up by adding well-rotted organic matter and grit but it is really important that you only work clay when it is dry. If you try to dig when clay is wet it becomes more solid and sticky.

Heavy clay soil should ideally be dug in autumn when soil conditions permit, incorporating lots of leafmould, and perhaps some grit, and leaving the frosts to help break up the clods into smaller crumbs. It's easy to tell if a clay soil is too wet to dig – your boots will be covered almost as soon as you start. If it is too dry it will set like concrete and you won't be able to get a spade into it. Even in optimum digging conditions clay soil can be heavy going, and you certainly won't want to dig it regularly. Once it has been dug over you should continue to open it up by planting green manure (see page 48) and regular mulching .

In theory, light sandy soil can be dug in any weather conditions as it does not compact like clay. But if you open it up by digging too much without adding copious amounts of organic matter sandy soil will become less rather than more nutrient retentive, losing moisture and organic matter. Never dig a sandy soil in autumn or leave it bare – cover it with a growing crop of green manure (see page 48) or a thick mulch of organic matter.

Digging methods

Neglected soils benefit from double digging (see page 32), where both subsoil and topsoil are thoroughly loosened. There are lots of methods of digging, but they all aim to do the same job: to aerate and fertilise the soil and leave it in a condition where it needs no further spadework.

If you dig to remove annual weeds the benefit is short-lived as you will bring nearer the surface the thousands of dormant weed seeds that live in the soil waiting for the right conditions to germinate.

Digging brings pests to the surface where they can provide meals for birds and other predators.

DIGGING METHODS

Instead of digging turf into the soil, you can remove it in about a 10cm layer and stack it, grass side down. If you water it and keep the pile covered with a polythene sheet you will have fine loam in a few months, ideal for potting compost or for adding to topsoil. But this seriously robs any bed where the turf came from, so it is best only to stack turf where you are removing it for a path or similar.

To decide your digging strategy, see what you're dealing with by digging a hole about two spade heads, or 'spits', deep. The soil will probably be rather compacted in the top spit but if it is reasonably loose underneath you can get away with light digging one spit deep, taking out weeds and forking organic matter in as you go.

Double digging

If the soil is compacted at the deeper level you need to loosen it for drainage and aeration or plant roots will never get the nutrients they need. The most successful way to do this is by double digging. Some people get quite fanatical about double digging but it is not difficult, just rather hard work. It involves turning over the subsoil as well as the topsoil, adding organic matter to each. The only rule is not to mix the levels of subsoil and topsoil.

The best way is to divide the plot to be dug into roughly 45cm squares. Take out a spit of topsoil from the first and second squares and a spit of subsoil from the first square, and leave these to one side. Then take the subsoil from the second square and turn it into the first square, adding compost or manure with it. Cover this with weeded topsoil from the third square, also mixed with organic matter, and so on. The topsoil and subsoil from your starting point ends up in the final squares.

Clearing grass

If you're clearing old turf you can't just turn it over or it will regrow and you'll have endless problems. Instead, use another digging method, called bastard trenching. As you do for double digging (see above), keep to your squares and dig two spits deep. Scalp a few centimetres of turf off the top of each square and bury it under soil in the neighbouring square.

TOOLS FOR THE JOB

Whichever tools you choose, make sure to keep them clean and sharp. Dirty tools can spread disease organisms into your soil and blunt tools make for hard work. Always clean your spade or fork after use by knocking off surplus soil and storing it in a sand box, or rub it with sand before hanging it on a wall. You can add a bit of old sump oil to the sand.

Don't overwinter a rotovator on a concrete floor – put it on a board or old pallet to avoid damp and rusting.

For excavating hard earth you may need to use a pick and shovel rather than a spade, although a spade is essential for cutting squares or trenches, and turning very heavy soil. In most other cases many experienced gardeners prefer to work their soil with a fork. As long as the soil is dry and not too heavy this is likely to be quicker than digging with a spade – a fork breaks up clods better, aerates the soil better and is easier to push into the ground. A fork is also better for incorporating organic matter, and the best implement for removing weeds.

If you're unlucky enough to take over a garden full of rampant creeping weeds such as bindweed, couch grass and ground elder you should only ever dig with a fork as a spade breaks up the roots into tiny pieces which become very hard to weed out and each one will sprout if you leave any behind.

Rotovating

When you want to cultivate a large area, or clear very weedy soil or rough grass, you may consider using a rotovator. Rotovators have tines that chop up the surface of the soil, and they can perform the equivalent of digging, forking and breaking down clods. However, you will need to rotovate a patch of ground several times to clear it as each time you rotovate you will chop up weed roots and potentially spread any that are not left on the surface to dry out. If you go over the ground a few times during the growing season when the plot greens up, you will eventually weaken the weeds and tough grasses, kill them and incorporate them into the soil.

Only ever rotovate soil when it is dry, to minimise weed spread. Unless you add generous quantities of organic matter to the soil as you clear it, even in the best conditions rotovating tends to harm the soil structure. If possible, sow a green manure with wide-reaching root spread after rotovating to help restore it (see page 48).

CULTIVATING WITHOUT DIGGING

The basic principle of the no-digging technique is always to leave a thick layer of well-rotted compost covering the surface of your soil, and to keep renewing it. You just plant your plants in that. Some people never dig their gardens at all. Not even to start. You can create beds without digging in all but heavy clay soil by covering the ground with layers of organic matter. Earthworms and other organisms should then incorporate it into the soil for you.

Don't try the no-dig method if you're starting a garden on unimproved heavy clay soil. Piling compost on top of cold wet clay would lead to your soil becoming increasingly airless, waterlogged and stale. The lack of aeration in heavy clay inhibits the organisms that break down the organic matter, so you could make a difficult soil worse.

Weed control

No-dig gardeners need to think ahead when planning their garden as the growing area will be out of action for at least one growing season, preferably longer, to clear the ground. The first step is to mow grass and chop down weeds on the area you want to turn into a productive bed, and water the ground well. Then spread a light-excluding layer of dampened newsprint (around 8 pages thick) or cardboard on the ground. This should prevent weed seeds germinating. Cover this with a 'sheet' of well-rotted organic matter, spreading a layer around 10cm thick to cover the ground completely. Cover this with another layer of moistened newsprint or cardboard to smother any weeds that get through the first layer and cover this with more compost. Keep everything damp so that living organisms can get to work rotting all the organic matter.

You should be able to plant into the new bed after six months, adding no more than 2cm of well-rotted compost and planting through that. If weeds appear, remove them with a handfork – don't disturb the bed by digging. Keep topping up with compost in spring and/or autumn.

THE MAGIC OF COMPOST

All plants and animals are made up of a complex range of proteins, minerals, carbohydrates, sugars and so on, so when you recycle organic matter into the soil it provides the ideal nutrients for plants and animals that feed on them to make more nutrients.

Don't worry if you can't make enough compost for your needs. You may be able to persuade neighbours to contribute organic material, but you can also buy a variety of readymade composts including recycled municipal waste (see page 44).

Good ingredients for compost include all kitchen scraps, weeds and garden waste, grass mowings and hedge clippings, manure, urine, hair, autumn leaves, pondweed, hay, straw, paper, teabags and coffee grounds, even old cotton socks and woollen sweaters.

Organic gardeners go on a lot about compost. That's because it's the best organic matter you can give to your garden. It's entirely natural, recycled from your waste products and free. Garden compost can provide your soil with all the nutrients and micro-nutrients it needs, it encourages helpful soil organisms and suppresses those that bring disease. You can even make it into liquid spray to combat fungal and viral diseases.

One of the principles of organic gardening is to try to maintain the fertile cycle of growth and decay by recycling nutrients, importing as little extra material as possible into your garden. The weeds and plants that go into a compost heap have spent their lives gathering minerals as well as nitrogen-rich proteins from your soil. When you return them to the compost heap all these elements are on their way back into the soil. And when you add the finished compost to your soil it then releases its plant foods slowly when they are needed.

Composting has been around for millennia in many different forms, and although you may come across compost buffs who like to imply that making compost is some arcane and complicated process, it is easy to make, and easy to use. If you simply collect a pile of organic waste in a heap, and keep it moist and aerated, it will gradually turn into a wonderful crumbly soil conditioner. The process will just happen – you can't stop it.

Why compost?
Whenever a plant or animal dies and falls onto the ground it will eventually rot down and be returned to the soil. This is part of the natural cycle of life. So why not just dig vegetation into the soil and leave it? There are several reasons why it is better to compost. One is to prevent temporary nitrogen depletion: the tiny organisms which rot fibrous vegetable matter by eating it use a lot of nitrogen to reproduce and grow in the process. So when they

are working to decompose matter in the soil or on the surface they are robbing the soil of nitrogen, starving growing plants. When they have finished their job they die and release the nitrogen they have used plus any from the vegetation. So you do get it back, but in the meantime you may have problems. Also, uncomposted material left lying on the soil surface may harbour pests and diseases whereas efficient composting will destroy them.

A compost heap is a purpose-built environment, a bacteria and fungus farm where the whole decomposing process works quickly and efficiently, mixing different materials to end up with a product containing a wide spread of balanced elements. If you just leave matter to rot on the soil you don't get this valuable mix.

There are dozens of different ways of making compost, and loads of useful ingredients to go with your kitchen and garden wastes. All methods should succeed as long as you follow a few rules. Use a mixture of soft and tougher materials, such as young weeds and strawy manure. Add as much diverse material as possible in one go. Never add more than about 10cm of grass mowings on their own. Keep your heap or bin moist and aerated and try to keep the heat in.

The best compost is rich, dark and crumbly. Useful compost can be produced in as little as six weeks, or it can take over a year – it depends on your method and materials. Your heap may never really warm up or it may get so hot that it steams. The compost may be smooth and crumbly or it may have lumps in it. Weed seeds may be killed, or they may not. It all depends what time of year it is, how much trouble you take and what materials and ingredients you include. Even if some weed seeds and lumpy bits do remain, it doesn't matter – you can either put the material through the composting process again, or use the compost as it is, it will still be useful, you'll just need to weed occasionally.

How you use your compost depends on your soil. Sandy soils contain a lot of air so any organic matter will oxidise and decompose quickly.

Clay soils are cold, wet and lacking in air, so garden compost must always be very well rotted and crumbly before use.

USING MANURE

Muck is full of micro-organisms, useful to feed and speed soil life and composting processes.

Never get manure from intensive rearing systems where extra minerals are often added to supplement the animals' indoor lifestyle and fast-food diet.

Don't worry about importing weed seeds in manure. As long as the animals didn't exist entirely on a diet of weeds, and the muck is rotted or composted, you are unlikely to have too much of a problem. Even if you do get a few weeds, it's easier to pull them out than it is to grow plants in weak infertile soil.

Never use manure fresh on your garden as the high levels of nitrogen can scorch plant roots and harm seed germination. It should always be rotted or composted before use. If you have a large garden make a muck heap in autumn, cover it with polythene and leave it to overwinter before adding to your soil. Otherwise simply leave muck to rot for a few months in plastic bags, tied at the neck to prevent air and water entering.

Manure has always been a valuable fertiliser. It provides high levels of nitrogen and also contains a good range of all the other nutrients your soil needs.

Choosing muck

Strawy manure is always best as much of the goodness in muck comes from the urine, and rotted straw is also a good soil conditioner.

Poultry manure is very rich – most useful when added to the compost heap to speed up composting rather than digging it straight into the soil. Well-rotted cow muck is an excellent fertiliser, and it is usually quite easy to come by, but town gardeners may find horse manure more available. This is also a good source of nutrients. But avoid muck mixed with wood shavings as the bacteria will spend all their time trying to decompose the sawdust rather than helping the fertility of your soil.

Pig muck is good, but be careful of its source – never get it from intensive-rearing systems as the pigs are often fed with copper supplements which can be toxic to your soil. Sheep and goat manure is valuable, but less available, and bat droppings are excellent, but hard to find.

Using muck

Only muck your soil thoroughly every three years or so as it is too nitrogen-rich for some plants and causes sappy growth and weak plants. It is best to apply muck to land about four to six weeks before you plant, or fork it carefully around growing plants, keeping it off their stems. Well-rotted muck and compost make a good mulch on clay soils, but not where you intend to make a spring seed bed.

Don't muck soils where you want to grow carrots or parsnips, or they will fang (split into several sections). Greedy feeders such as potatoes love muck, but it must always be very well-rotted as immature muck encourages potato scab.

HOW AND WHEN TO MULCH

Gardeners should steer away from peat because of the environmental problems in digging it, but even if you live in an area of plentiful peat, never use it as a mulch. It forms a water-repellent crust that sheds instead of absorbs rain, and if it is left on top of the soil it takes an age to break down. So peat is a poor source of organic matter and nutrients.

If you want to make a seed bed in spring on clay soil mulch it with well rotted compost or leafmould over the winter, and rake off excess in spring. You won't be able to dig a mulch into clay soil until late in spring because of the difficulty in cultivating clay soils except in dry weather when the soil is not wet.

If you want to improve and protect your soil's fertility, mulch. Mulching means covering the ground with a layer of material (see page 36). It is the easiest way of adding organic matter – just spread the mulch on top of the ground and let worms and other soil creatures do the work of incorporating it into the soil. This will also suppress most weeds by keeping light out.

Why mulch?

Mulching keeps the soil moist and prevents it drying out in hot weather; it also prevents soil getting waterlogged in winter. A layer of mulch shades soil, reducing heat stress on plants and soil organisms. It helps the soil temperature to stay reasonably constant, keeps it warmer for longer in autumn to extend the growing season, and insulates it in winter. This regulating effect helps soil structure, plants and soil life.

When you leave unplanted soil well mulched over winter this prevents erosion from extreme weather, and also helps maintain the soil structure.

Mulching materials

The most common mulches are organic materials. Compost, well-rotted strawy muck, grass mowings or leafmould may be applied as slow release fertilising mulches. These can also be used to add a protective layer to the soil over winter. If you're short of compost, wetted straw and hay also make good winter protection – a small proportion will get worked into the soil but rake off the excess in spring to avoid starving the soil of nitrogen.

Woodchips can make an attractive mulch, but leave them in a pile to decay for six months to a year before using or they can tie up enough soil nitrogen to interfere with plant growth. Grit makes an effective mulch for plants that require good drainage, and to keep slugs and snails at bay, but don't use it on sandy soils.

Mulching weeds

You can use any organic matter for a sheet mulch to keep weeds at bay, or combine it with sheets of newsprint or cardboard (see page 36). Some gardeners use black plastic or carpet for this purpose. The best inorganic material is porous horticultural plastic, which stops weeds effectively, but doesn't disable soil life. Natural woollen carpet is also effective. It eventually rots into the soil to provide more nutrients, but be careful using nonporous black plastic which smothers soil organisms along with the weeds.

Whatever sheet method you choose, always improve your soil properly before laying the sheet mulch; once a mulch is in place it's hard to get at the soil to replenish soil life. You can plant shrubs, trees, perennials and most vegetable seedlings through carpet, cardboard and porous plastic.

When to mulch

Your soil must be fairly wet before you add mulch, as rainwater will only trickle slowly through the mulching material and some will be absorbed before it gets to the soil. On the other hand, your soil won't lose much moisture and even in a hot dry summer it should stay moist a few centimetres beneath the mulch.

Unless you are following the no-dig method, in spring you should fork in or rake off excess winter mulch a few weeks before planting to let soil warm up, and wait to re-mulch perennial plantings until plants put out new growth. Then mulch well to ensure that soil stays fertile and moist through the summer. In autumn it can pay to rake off excess summer mulch and let birds get at the grubs and pests in the soil. Wait until the first hard frost before you re-cover beds with a thick layer of loose mulch to prevent winter erosion and as good insulation from the thawing and freezing that harms plants. Straw, hay and chopped bracken make good insulating mulches.

Don't mulch very wet soils as this can promote some diseases and increase chances of rot. If spread too thickly mulches can interfere with the soil's aeration.

Leave soil bare for a few weeks in autumn before spreading winter mulch. This gives birds the chance to dig through the soil and eat pests and pests' eggs and larvae.

One disadvantage of mulch is its other probable role as slug and snail hotel, providing the warm moist conditions they love. So be vigilant, especially when you rake off a temporary mulch. On the plus side, it is possible that mulch may also provide slugs and snails with an alternative food source to your plants.

MORE SOIL IMPROVERS

If you have a large amount of autumn leaves don't add them to the compost heap but rot them separately.

In some areas recycled municipal waste is available. Although this is of variable quality, it can be a good source of bulky organic matter, and far better to get garden supplements from a local source than imported. Some local authorities compost grass mowings, many have schemes to recycle shredded prunings. Others provide composted humanure. Ask about the content of what is available.

Peat is no longer recommended because of concerns about the environmental impact of its extraction.

Few gardeners generate enough compost from their own-household and garden to keep their soil in best condition. Fortunately plenty of other materials are widely available.

Grass clippings

Lawn mowings are best left on the lawn, or composted, but they also make a good mulch on sandy soils, and an excellent nitrogen boost for potatoes if you lay them straight into potato trenches. Never leave a thick layer of fresh grass clippings on top of clay soil as they will rob the soil of nitrogen while they decompose, or turn to slime and further prevent aeration of the soil.

Leafmould

Autumn leafmould is not a fertiliser as it contains scarcely any plant foods, but it adds bulky organic matter to condition soil, and makes excellent weed suppressant mulch.

The easiest way to collect autumn leaves from a lawn is to run the mower over them with the grassbox on. This chops the leaves and mixes them with grass clippings which will speed their decay. Otherwise rake them up when they are damp and either pile them into a heap and cover it with polythene, or put small quantities into loosely tied plastic sacks. Leave them for at least six months before digging them into your soil or using as mulch.

Seaweed

The unrestricted use of seaweed, either fresh or calcified, to reduce soil acidity is not recommended by the Soil Association. Taking fresh seaweed from the shore damages the natural ecosystem of the beach and endangers certain seaweed species. Calcified seaweed is a restricted product for organic growers and farmers. Ground limestone or chalk should be used as an alternative.

Comfrey

Everyone should grow comfrey as a mineral-rich fertiliser. This deep rooted plant is low in fibre and high in protein, the best source of potassium for organic gardeners, and contains significant levels of nitrogen and phosphorus. It grows faster than any other plant in your garden so you can crop stems and leaves every six weeks through the growing season. No pests will attack your comfrey plants.

Cut comfrey stems and leaves and wilt them for one or two days before digging them straight into the soil. Comfrey also makes an excellent summer mulch around tomatoes and bush fruit, adding fertility and deterring pests and diseases. However, don't use comfrey on acid-loving plants.

Mushroom compost

Organic mushroom compost is an excellent soil conditioner, fertiliser and mulch, as long as it is based on well-rotted manure. However, much commercial mushroom compost contains pesticide residues as well as high levels of lime. Leave it in a pile under cover for six months before using it, and don't use it on acid-loving plants.

Other conditioners

Composted green bracken adds phosporus and potassium. Harvest it in summer, don't wait until bracken dies down in autumn as it is then low in nutrients, although you can use this to insulate tender plants.

Hoof and horn provides slow release of nitrogen, bone-meal provides calcium and phosphorus, both are good general fertilisers to build strong root growth but check there are no chemical additives used in the sterilising process. Blood, fish and bone also makes a good spring tonic to get growth off to a good start. It releases nitro-gen fast and other elements, including trace elements, more slowly.

The best comfrey for gardeners is a variety called Bocking 14, selected by HDRA from varieties imported from Russia. It is a hybrid of common comfrey (*Symphytum officinale*) and prickly comfrey (*Symphytum asperum*) and is sometimes known as Russian comfrey. Comfrey can be used as a quick-rotting mulch, but don't use it on flowering plants.

Hair, fur, wool and feathers all release nitrogen slowly. Either compost or dig them into the soil in autumn.

Woodash provides potash, but also salt. It is best not to add woodash directly on the soil but add it to the compost heap. Woodash is highly soluble but nutrients are incorporated in compost in a less soluble form.

LIQUID FERTILISERS

Compost tea
Make this like liquid manure, or simply fill half a bucket with compost and top up with water, cover it and leave to steep for two weeks before diluting and using.

Comfrey liquid is excellent for tomatoes, peppers, aubergines, squashes and marrows, peas, beans and soft fruit bushes.

Comfrey and nettle liquid are slightly alkaline and so are not recommended for acid-loving plants.

Organic gardeners should always try and 'feed the soil, not the plants' but there is also a place for liquid fertilisers. They feed plants directly, providing them with readily accessible nutrients, and are most useful to boost plants while you are still building your soil's fertility. They also act as foliar feeds where roots are restricted, such as when plants are growing in containers.

Comfrey liquid
The most popular use of comfrey (see page 45) is in compost or as high potash liquid manure. If you have the space, place about 8 kilos of chopped comfrey in a 90-litre water butt, fill to the brim with water and cover tightly. In about four weeks a clear liquid can be drawn off. Use this without diluting it as a general pick-me-up for potash loving plants, or for anything that needs a tonic, even houseplants. Or you can make small batches of comfrey liquid in a plastic bucket. A few gardeners recommend using urine in place of some water for even higher nitrogen levels. Always cover a tub or bucket of comfrey liquid as the brew smells awful and also attracts flies and mosquitoes.

Nettles and horsetail
Made in the same way as comfrey liquid, nettle leaf liquid makes a good general feed, high in major elements and magnesium, sulphur and iron. Horsetail is high in all soil nutrients, and rich in silica.
    Dilute these liquids about 10 to 1 and use them directly onto plants or to activate compost.

Liquid manure
Fill a hessian sack with farmyard muck and suspend it in a barrel of water. After two weeks the liquid is ready to use. Liquid manure is a powerful plant food – never take it internally, use it sparingly on crops and avoid spraying on the edible parts of plants.

GREEN MANURES

Alfalfa is a perennial green manure that can be left in the ground for over a year, and cut down for composting two or three times.

Clovers suitable for green manuring are best on light soils, performing poorly on clay.

Many summer green manures also attract beneficial insects into your gardens. Buckwheat is an excellent summer weed suppressor and bulky green manure. It will grow on very poor soil, and hoverflies love it.

Phacelia (*Phacelia tanacetifolia*) is an attractive summer green manure, ideal for the garden with lavender blue flowers that bees and other insects love. It will also withstand a mild winter.

Green manures are fast-growing crops that you grow for the primary purpose of turning them into the soil to add organic matter. They are sometimes known as cover crops or living compost. When the crop is mature, and before it flowers, you dig it in to the ground or cut it and leave it on top of the soil as a mulch for the worms to incorporate into your soil.

Green manures improve soils in several ways. They prevent weeds colonising bare soil; they increase biological life in the soil; they help to prevent pests and diseases from establishing themselves; and they add bulky organic matter.

Even a small patch of green manure is effective. It is most often used in the vegetable patch, both as a way of improving the soil when you start your garden and as part of your rotation plan. If you leave an area of soil bare, nutrients in the soil will be carried away by wind, rain or even heat, nitrogen will be lost in ammonia gas, and harsh weather can also damage soil structure. But if a green manure crop is growing, plant moisture and nourishment will stay in the soil, and its structure will be improved rather than eroded by weather conditions.

In a small garden you're most likely to want a winter growing green manure to cover resting soil. In a larger garden summer green manures are useful if you don't want to cultivate the whole area, particularly to suppress weeds and bolster fertility as part of your rotation system (see page 52).

Adding nutrition, improving structure

Legumes and grasses are the most common green manures. Legumes includes peas and beans, clovers and alfalfa. They can fix nitrogen from the air in their root nodules and return it to the soil. Turning legumes into the soil at any growth stage adds organic matter and improves soil life.

Grasses provide dense cover with good root penetration; they are particularly valuable for weed control and to improve structure. Many grasses have extensive root systems. When the roots decay they leave not only organic matter but also hundreds of fine channels in the soil. So they aerate the soil and improve drainage.

Some legumes, notably alfalfa but also clovers, have deep root systems, reaching right down to take nutrients from the subsoil. When the plants decompose these nutrients are returned to the topsoil where even shallow rooted plants can use them.

Winter cover

Winter tares are one of the few hardy legumes. When sown in late summer they will grow until well into the winter and stand until spring when you can cut them down and dig them in as nitrogen-rich compost. Phacelia and crimson clover will also survive mild winters.

All cover crops suppress weed growth, and grazing rye also acts as a natural herbicide producing toxins from its roots that kill many weed seeds and seedlings, even couch grass. However, when the foliage is turned into the soil this can prevent small seeds germinating, so follow rye with large seeded vegetables such as beans, and wait to sow until three to four weeks after digging in the plants.

Specific strategies

If you are turning old pasture land into a productive garden, grow mustard before planting vegetables. Grassland usually harbours wireworms, the larvae of the click beetle. They feed on grass roots, so when you clear the grass they will feed on whatever comes next, particularly potatoes and other root crops. Luckily mustard is a favourite food and when you turn it under the soil in spring the larvae feed on it so greedily that they become beetles in record time and fly away to lay eggs in grassland elsewhere.

Although not strictly a green manure crop, if you have a severe problem with soil pests and perennial weeds, try planting marigolds (*Tagetes minuta*). These are a bit tough for digging straight in but make excellent bulk for the compost heap.

Don't overwinter green manure on heavy clay if you're planning a spring seedbed as conditions may not be favourable for digging it in at the time you want.

DRAINAGE DIFFICULTIES

In a very heavy soil, you may not consider drain building is the best way of spending time and energy. Instead, build raised beds (see page 56) extending at least 15cm above the soil surface. Double dig the soil first.

Peat soils tend to get soggy and waterlogged and need draining. Adding organic matter will not improve these soils.

Drainage problems may be caused by poor soil structure and can be fixed by adding organic matter and building up the fertility of your soil. But sometimes you may need to find other solutions.

Poor drainage

When the water table is near the surface or the subsoil is heavy clay or rock, soil will get waterlogged, plants and soil-living organisms will not be able to get enough oxygen and soil will stagnate. If double digging, green manuring and adding organic matter fail to solve the problem you could take advantage of the conditions to create a bog garden. But if you want a wider choice of plants you will have to make drains.

Rubble drains are the simplest do-it-yourself drains for a small plot. Dig a trench 60-90cm deep and 30cm wide, with a gentle slope of about 1 in 30, preferably leading to a ditch or drain. Half fill it with rubble, top with gravel and replace the topsoil. If there is nowhere for your drain to go you will need to dig a soakaway, a hole at least 1.5m wide and deep, lined with bricks, filled with rubble and topped with turf. You should dig your drainage trenches by hand as even a small excavator will further compact your soil and add to fertility problems.

Excessive drainage

Some very thin stony soils, and those on steep slopes, will not retain water, and it would take lifetimes of adding organic matter to make much difference. Never leave areas of thin soil bare but keep covered with green manures or groundcovers and build raised beds for productive gardening, importing topsoil and organic matter. Steep slopes may need to be terraced in some way to create level areas where water can seep in rather than just running off, taking topsoil and nutrients with it. Or choose plants that tolerate arid conditions.

ROTATION

If bed space is limited, ease difficulties with rotation and grow some vegetables in containers.

Leafy salad vegetables and other swift-growing shallow-rooted vegetables can be grown between other crops.

Never grow tomatoes in the same place where potatoes have grown the previous year as they are members of the same family.

You should try to avoid growing the same vegetables in the same positions in your garden year after year, as this can lead to problems with soil pests, diseases and fertility.

Pests and diseases

Crop rotation is a practice from agriculture that should also be followed as far as possible in a small garden. The standard technique is to use a four-year rotation to avoid the build up of crop-specific soil pests and diseases. For example, onion white rot will build up to attack onions, garlic and leeks, eelworms attack peas, beans, onions and potatoes, and clubroot can harm all brassicas. By rotating these crops you help avoid continuing problems.

Some people say that rotation makes little difference in a small garden, as many pests and diseases stay in the soil for much longer than four years and you can only move crops a few metres. But at the very least it is still a valuable way of improving your soil.

Maintaining soil fertility

Different crops take different things from the soil, so if you grow one species continually this can exhaust the soil of a particular range of nutrients. But if you change crops each year you replenish the soil with different elements. Peas and beans, for example, are useful in a crop rotation system to provide the soil with nitrogen to feed a following crop.

Rotation also helps your soil structure, particularly when you grow a range of shallow-rooting and deep-rooting plants.

Weed control

Rotation helps weed control. Greedy feeders and bushy plants such as potatoes and squash, for example, are good at clearing ground for a following crop. Enzymes from tomato roots help inhibit growth of couch grass.

Vegetable rotation systems

A traditional system rotates potatoes, brassicas, legumes and roots on a four-year cycle, but few home gardeners follow this slavishly. First of all think about what you want to grow and eat. There's no point in growing something you don't want just because it fits into a fertility system. Then think about the different needs of specific plants. For example, leafy greens and brassicas also thrive in soil where beans have fixed plenty of nitrogen; onions like well-fed soil but carrots and parsnips will 'fork' in too-rich conditions; potatoes don't like lime.

You also need to think of sowing and harvesting times. You want to avoid treading on your soil wherever possible, and you certainly don't want to be walking around growing crops so it's most convenient to have adjoining areas of your plot clear at the same time.

Try to leave room for regular green manure crops for cover and structure, and remember to fit them into the rotation system – don't follow cabbages with mustard, for example, as they are both brassicas. Try to get into the habit of planning carefully, and try to keep good records of what you plant when, and when you harvest. But you'll learn most by your successes, and by any failures.

Ignoring rotation

Sometimes it may not be practical to follow a rotation system – for example, only one part of your garden may be sunny or warm enough for certain crops, or in a very small garden you may want to concentrate on quick-growing vegetables or summer salads. As long as you maintain your soil in good condition, keep adding plenty of organic matter and don't grow exactly the same thing in the same place two years running, then you should safely be able to grow most things without too much worry. Just keep an eye out for any problems and never follow an infested or diseased crop with the same species.

Vegetable families

• **Brassicas**
Brussels sprouts, cabbages, cauliflowers, kale, radishes, swedes, turnips, mustard

• **Legumes**
Peas and beans

• **Potato family**
Potatoes, tomatoes, aubergines, peppers

• **Umbellifers**
Carrots, parsnips, celery, parsley, fennel

• **Daisy family**
Lettuce, chicory, endive, scorzonera

• **Onion family**
Onions, garlic, leeks, chives

• **Beetroot family**
Beetroot, spinach, chard

• **Cucurbit family**
Courgettes, marrows, squash, cucumbers

BEGINNING IN BEDS

A bed system is versatile and easy to manage – for example, you can manure one bed each year for greedy feeders, or lime one bed at a time, creating slightly different soil environments for specific plants.

If your beds are oval or rectangular, try and lay them north-south so that you can plant taller plants at the north end to minimise their shading effect.

Fertile soil can't be made out of poor soil overnight – it takes time. Time for the gardener to build it up with good organic practice and time for the soil to adjust to the new developments. But it's also natural to be impatient. However, the key to successful gardening is never to take on too much at one go. Otherwise jobs get left half done or don't even get started, and there is a constant pressure that can turn gardening from a pleasure into a nightmare.

A sensible first step, particularly for vegetable growers, is to divide your plot into manageable beds, separated by paths. No-dig gardeners usually use beds, and it is an equally useful soil management strategy for conventional dig-and-mulch gardeners, even if you have a large garden with a substantial area dedicated to fruit and vegetables.

Growing in beds means you can concentrate your resources on the areas that need them. It is much more productive, for example, to spread organic matter thickly over a dedicated area, rather than a blanket approach. The fertility of a few small areas can be built up quickly compared to tackling a large area.

Ideal beds

The best width for beds is around 1.5m so that you can reach the centre of a bed without ever treading on the soil. Paths should be wide enough to take a wheelbarrow. All weeding, cultivating and planting should be done from the paths so you will never compact the soil and harm the structure by treading on it.

Even difficult soil in beds can be worked at most times of the year because there is no fear of compaction when cultivating. You can also plant closer together in beds than in more open ground, which means you have less weed problems but grow more produce. If you make a minimum of four beds, however small, this makes rotation (see page 52) simple to manage, further helping to maintain your soil health and fertility.

RAISED BEDS

It is always best to grow plants that will love your soil. They will thrive with minimal attention. If your soil is lime-rich, but you can't resist growing a few acid-loving plants, for example, you could build a deep raised bed. But be aware that this will still require more maintenance than other areas of your garden where organisms will be adapted to a more alkaline environment.

It's not always possible to plant straight into your soil. You may take over a derelict garden that needs time and work before the soil will be in a condition where it can support healthy plants. Or your topsoil may be very thin, or drainage could be bad. Some new developments on brownfield sites may even have a problem with contaminated soil. Or you may have mobility problems and be unable to work the soil at ground level comfortably. You can still get growing, using raised beds.

Making raised beds

Beds can be as high as you find comfortable. For best results, double dig the area first, or at least aerate the subsoil thoroughly with a fork. Make sure your raised beds are no more than two arms' reach in breadth for ease of working, and mark them out with an edging of your choice. Wooden boards are practical but avoid those treated with chemicals, or use bricks, tiles, stones, depending on what is available locally. You will then need to fill the bed to the required depth with a mixture of topsoil, compost and other organic matter.

Unless you are using loam made from your own turf-stacks, watch out for the quality of imported topsoil – it can be very sterile and full of weed seeds. Ideally, make raised beds in autumn and sheet mulch with cardboard covered with a layer of organic matter through which you can plant in spring. If you need to make higher raised beds, make sure the edging is sturdy enough to take the large volume of soil and compost.

Raised beds for added fertility

You can grow plants more closely together in deep raised beds than in ground-level planting because of the depth of organic matter. You will never need to dig, and soil structure and fertility should maintain themselves with nothing more than regular addition of compost.

SOIL IN CONTAINERS

Container soil should have a loose crumbly texture to absorb nutrients but retain air and drain easily. It must contain plenty of food supplies. Use a mixture of garden loam or molehill soil and well-rotted leafmould for the base, adding blood, fishmeal, bonemeal, worm compost or well-rotted garden compost for nutrients, and sand or grit for extra aeration.

Be careful when sterilising soil – it is easy to overbake it and the process smells.

Garden compost is a fertiliser and soil conditioner, and can't be used on its own as potting compost. Worm compost, however, makes ideal potting compost.

Garden soil should only be used as a base for permanent plantings in large containers, such as fruit trees or large shrubs.

Foliar feed your container-grown plants with liquid fertilisers (see page 46).

Where space is limited, many plants can be grown successfuly in containers but sometimes even the most experienced gardeners have difficulty growing in pots. Just as in the garden, the secret is the soil. It should be light and loose for drainage and air, but not so light that nutrients drain straight away. While soil structure and fertility in the garden is maintained via all the living creatures and micro-organisms, you can't create a living soil in a container, so you have to follow different rules.

Choosing soil

Garden soil is a poor growing medium for plants in containers. Soil compacts much more easily in a pot than in the garden, quickly becoming heavy and airless. Every time you water the soil tends to settle, and there are no living organisms to stir it up again. Container soil therefore needs to be much looser than garden soil. You can use a proprietary potting compost, or make your own from loam, well-rotted composts and sand or grit – then you know exactly what's in it.

The best soil base for potting compost is loam from turfstacks, or molehills where soil is weedfree and well broken down into crumbs. If you use garden soil as a base, sterilise it first to kill any disease organisms and weed seeds by baking it at around 180° C in the oven for 30 minutes.

Container size

Avoid stress on container-grown plants by making sure their pots are the right size for them. A plant will tell you when it needs repotting by wilting easily and sending roots out through the container's drainage holes. Always repot into a container that is only one size larger than the one your plant has grown out of. Soil in a big pot containing a small plant will always stay wet so the plant runs the risk of root rot and other stress.

RESOURCES

Organisations to join
Centre for Alternative Technology (CAT)
Machynlleth
Powys SY20 9AZ
01654 702400
info@cat.org.uk

HDRA, the organic organisation
Ryton Gardens
Ryton on Dunsmore
Coventry
Warwickshire CV8 3LG
024 7630 3517
enquiry@hdra.org.uk

Soil Association
Bristol House
40-56 Victoria Street
Bristol BS1 6BY
0117 929 0661
info@soilassociation.org

Mail order
Chase Organics
Riverdene Business Park
Molesey Road
Horsham
Surrey KT12 4RG
01932 253666
www.OrganicCatalog.com
*suppliers of organic seeds and gardening tools
and equipment, including compost bins*

Gardening on the web
Centre for Alternative Technology (CAT)
www.cat.org.uk

HDRA, the organic organisation
www.hdra.org.uk

Organic UK
www.organic.mcmail.com

Soil Association
www.soilassociation.org

The Compost Resource Page
www.oldgrowth.org/compost

More books to read
Peter Harper, *Natural Garden Book*, Gaia
Books, 1994

Lawrence D. Hills, *Fertility Gardening*, David
and Charles, 1981

John Seymour, *The Complete Book of Self
Sufficiency*, Dorling Kindersley, 1997

Elizabeth P Stell, *Secrets to Great Soil*, Storey
Books, 1998

Sue Stickland, *The Small Ecological Garden*,
Search press/HDRA, 1996

INDEX

acid soils 22, 23, 26; plant choices 24-5
aeration 13, 19, 26, 31, 32, 36, 39, 43, 44, 49, 56, 58
air 12, 13, 18, 19, 30, 58
alfalfa 48, 49
alkaline soils 22, 23, 26, 46; plant choices 24-5
ammonia 23, 48
animal/farmyard manure 18-19, 40, 42

bacteria 12, 22, 40, 44
bastard trenching 32
bedrock 12
beds 36, 54; raised 14, 23, 50, 56
bracken 45

calcium 20, 45
carbohydrates 12
carbon dioxide 13, 20
checklists: planting 24-5; soil basics 26-7
clay (heavy) soils 16, 19, 20, 26-7, 30, 36, 45, 49, 50; improving 22, 23, 26, 31, 39; mulching 39, 40, 42
clovers 48, 49
comfrey 20, 27, 45, 46
compaction 14, 31, 54, 58
composts/composting 22, 27, 36, 39, 42, 44, 58; preparation and materials 18, 38-9, 44, 45
container growing 46, 52, 58
crop rotation 52-3
cultivation 26, 27, 36, 42, 54

decomposition 39, 40, 44
digging 19, 30-2, 36, 49
diseases 13, 34, 38, 39, 40, 43, 52
drainage 14, 19, 26, 27, 32, 49, 50, 58

earthworms 23, 30, 36
erosion 22, 42

fertilisers 40, 45; artificial 9; liquid 46, 58
fertility 13, 18-19, 30, 31, 38, 46, 52, 56, 58
foliar feeds 46, 58
fungi/fungicides 12

garden compost 39, 58
grass mowings 39, 42, 44
grasses 48-9
grass/ground clearance 30, 32
gravel/grit 13, 19, 26, 31, 42
green manure 19, 31, 34, 48, 53
gypsum 23, 27

hardpan 14, 30
herbicides 49
humus 12

insects 12, 13, 48
insulation 42, 43, 45
intensive rearing systems 40
iron deficiency 27

leaching 18, 26
leaves/leafmould 20, 42, 44, 58
legumes 48, 49, 52, 53
life-cycle 13, 38
lime/liming 14, 22, 23, 26, 27, 45, 53, 54
loam soils 16, 19, 26, 27, 58

magnesium 20, 27
management/strategies 34, 36, 49, 54; seasonal variations 36, 39, 42, 43, 45, 48, 49
manganese deficiency 27
manures 20, 39, 40; animal 18-19, 40, 42; green 19, 31, 34, 48, 53; liquid 46
marigolds 49
micro-organisms 18, 40, 58
minerals 12, 18, 20, 22, 27, 38
moss 22

muck *see manures*
mulching 19, 30, 31, 36, 40, 42-3, 45, 48
mushroom compost 45

newsprint/cardboard, use of 36, 43
nitrogen (N) 20, 22-3, 39, 40, 44, 45, 48, 52;
deficiencies 20, 27, 38, 42
no-digging cultivation 36, 43, 54
NPK balance 20
nutrients/nutrition 12, 14, 20, 22, 26, 30, 38, 48;
deficiencies 20, 27, 38, 42

organic matter 12, 18-19, 30, 36, 38, 44
organisms (soil) 13, 18, 19, 22, 31, 36
oxygen 13, 14, 44

peat: as compost 42, 44; soils 12, 16, 26, 27, 50
pests 13, 31, 39, 43, 49, 52; control 45
pH levels 20, 22-3, 26, 27
Phacelia 48, 49
phosphorus (P) 20, 27, 45
plants/planting 22, 38, 43, 54, 58;
needs/nutrition 8, 12-13, 18; problems 20, 27;
selection 14, 24-5, 56
potash 20, 46
potassium (K) 20, 27, 44, 45
potting compost 58
poultry manure 40
problem solving 22, 49, 50
proteins 12, 38, 45
puddling 18, 19

recycled municipal waste 38, 44
rock dust 20, 27
root penetration 30, 49
rotation 52-3, 54
rotovators/rotovating 34
rubble drains 50

sandy (light) soils 16, 19, 23, 26-7, 30, 31, 42;
improving 26, 39, 44, 48
sea sand 14
seaweed 44
seed beds/seedlings 39, 40, 42, 44
silt soils 13, 26, 27
slugs and snails 42, 43
soils: analysis/testing 16, 20, 22; checklists
26-7; content 8-9, 12-13, 14, 58;
improvement 19, 22-3, 28, 30, 44-5, 48;
problem solving 22, 27; structure/texture
13, 18-19, 26-7, 31, 34, 42, 48, 58; *see also clay
soils; fertility; sandy soils*
sterilising 45, 58
straw/strawy manure 40, 42
subsoil 12, 14, 32, 56
sulphur 20, 23, 27

temperatures: soil 26, 27, 42
tools 34
topsoil 12, 14, 30, 32
trace elements 20, 23, 44, 45
trenches/trenching 23, 32, 50
turf 32

urine 40, 46

vegetables 22, 23, 40, 44, 52; families 53

waste materials for composting 38, 39, 44
water/waterholding 12, 13, 18, 26, 30
waterlogging 19, 26, 36
weeds 22, 31, 34, 38, 39; controls/mulching 36,
42, 43, 44, 48, 49, 52, 54
woodshavings/woodchips 40, 42
woodash 45
worms 12, 23, 30, 36, 49

A GAIA ORIGINAL

Books from Gaia celebrate the vision of Gaia, the self-sustaining living Earth, and seek to help its readers live in greater personal and planetary harmony.

Design	Lucy Guenot, Mark Epton
Editor	Pip Morgan
Index	Mary Warren
Photography	Steve Teague
Production	Lyn Kirby
Direction	Joss Pearson, Patrick Nugent

This is a Registered Trade Mark of Gaia Books Limited.

Copyright © 2001 Gaia Books Limited, London
Text copyright © 2001 Charlie Ryrie

First published in the United Kingdom in 2001 by
Gaia Books Ltd, 66 Charlotte Street, London W1T 4QE

ISBN 1 85675 122 8

A catalogue record of this book is available from the British Library.

Printed and bound in Italy

10 9 8 7 6 5 4 3 2

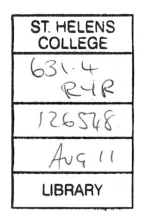

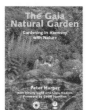